Twist, Wiggle, and Squirm

A Book about Earthworms

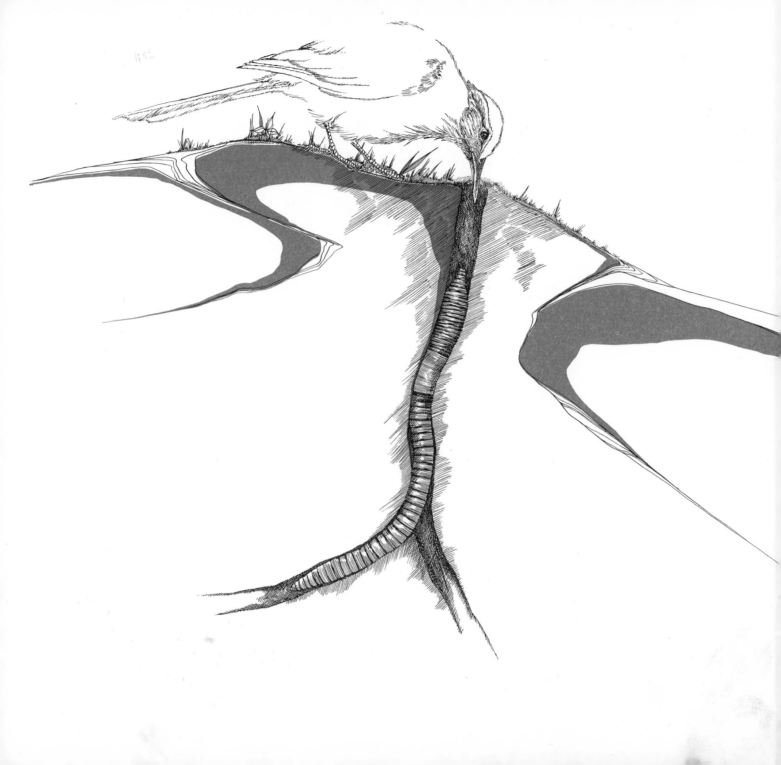

Twist, Wiggle, and Squirm

A Book about Earthworms

BY LAURENCE PRINGLE

Illustrated by Peter Parnall

Thomas Y. Crowell Company   New York

LET'S-READ-AND-FIND-OUT SCIENCE BOOKS

Editors: *DR. ROMA GANS*, Professor Emeritus of Childhood Education, Teachers College, Columbia University
DR. FRANKLYN M. BRANLEY, Astronomer Emeritus and former Chairman of The American Museum-Hayden Planetarium

**Available in Spanish*

L.C. Card 74-184983

3   4   5   6   7   8   9   10

Twist, Wiggle, and Squirm

A Book about Earthworms

LET'S READ AND FIND OUT

Earthworms squirm.
They twist and turn.
They crawl and burrow in the soil.

After a heavy rain you may find hundreds of earth-
worms on lawns and sidewalks.
Perhaps you use worms as bait when you go fishing.
There are about 2,700 different kinds of earthworms.
Some live in lakes and ponds.
Most of them live in the soil.

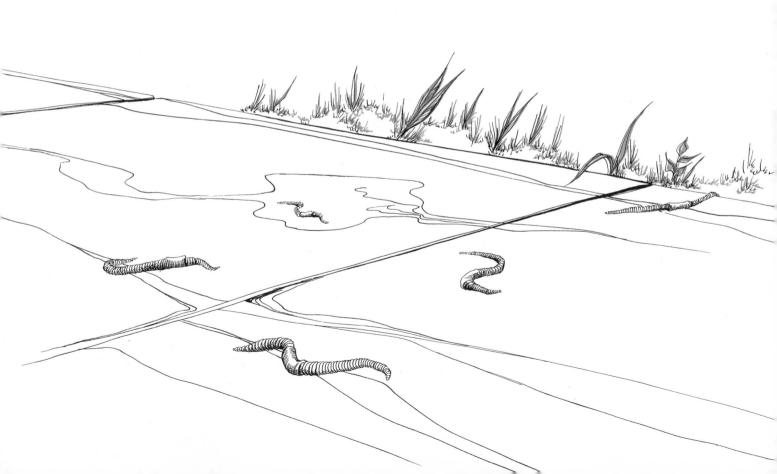

Some kinds of earthworms are tiny.

But most of them are a few inches long.

There is one kind of worm in South America that grows to be seven feet long.

The world's biggest worms live in Gippsland, Australia.

Sometimes they are twelve feet long, and weigh a pound and a half.

These giant worms make gurgling sounds as they move under the ground.

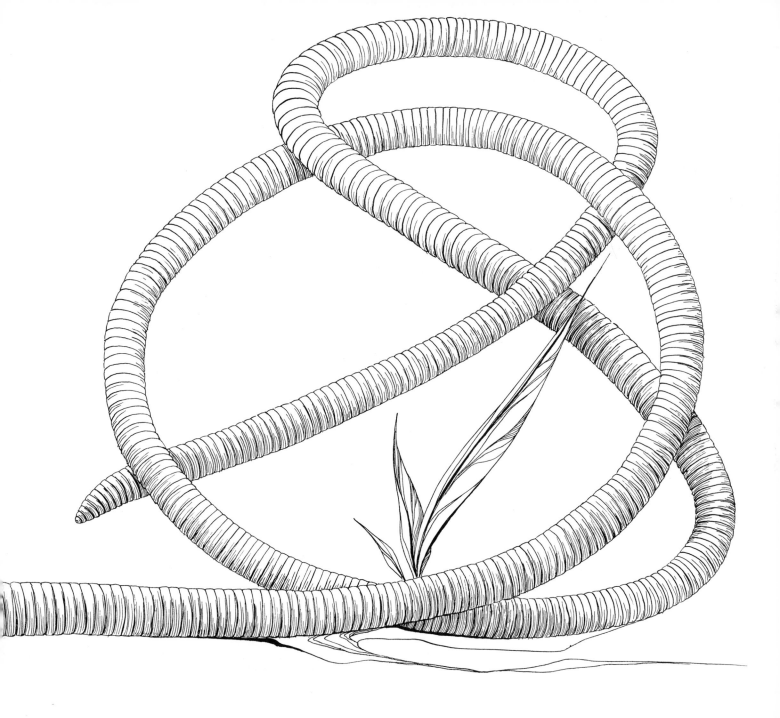

3

All earthworms, big or little, are alike in some ways.

They have no bones.

Their bodies are made up of many rings, or segments,
of muscle.

These segments are held together by other muscles.

Earthworms can twist into many shapes.

They can even twist into a knot.

They can stretch out long and thin.

They can squeeze their segments together, and make
themselves short and fat.

Earthworms have no legs.

They move by stretching and squeezing.

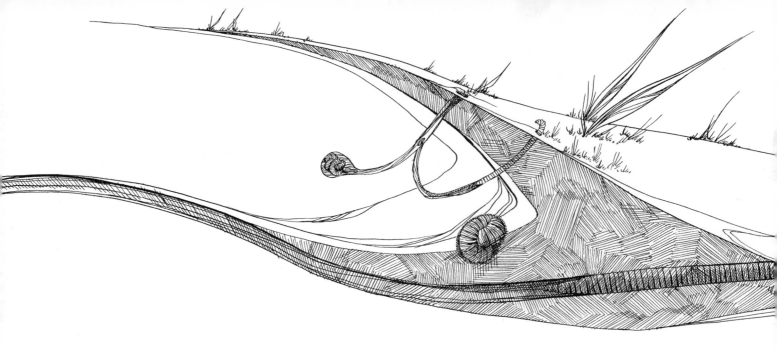

Often there are several kinds of worms living in the
soil. These may be garden worms, red worms, dug
worms, leaf worms, and swamp worms.

One common kind of earthworm is called the night-
crawler.

It sometimes grows to be a foot long.

Nightcrawlers have about 150 segments in their
bodies.

Like all earthworms, nightcrawlers are most active
at night.

During the day they stay down in their tunnels.
At night they crawl out to feed and sometimes to
  mate.
Fishermen use nightcrawlers for bait.
They catch the worms at night.
You can catch them too.
Take your flashlight and go out on a warm, still
  night when the soil is wet.
In lawns, fields, and gardens, nightcrawlers will be
  stretched partway out of their burrows.

Walk softly.

Earthworms do not hear sounds the way people do,
but they can feel the ground shaken by heavy foot-
steps.

If you step gently, they will stay on the surface.

Shine the flashlight around and look for some nightcrawlers.

Move quietly close to one and shine the light on it.

Zip! It slips into its hole.

Earthworms have no eyes, but they can sense the light on their bodies.

You can get close to a worm if you don't shine the light right on it.

You can also get close if you cover the glass front of a flashlight with thin red paper or plastic.

Worms do not sense red light.

You may have seen a robin trying to pull a worm from the ground.

Sometimes the robin tugs and tugs but the worm will not budge.

You can find out why by catching a worm yourself.

Pull the worm slowly back and forth through your fingers.

You will feel tough little bristles on each segment.

They are called setae.

Earthworms have four pairs of setae on nearly all of their segments—two on the bottom and two on the sides.

The setae help earthworms grip the insides of holes in the ground.

They also help the worms crawl.

The setae hold the soil as the worm stretches and squeezes the segments of its body and crawls along.

If you tug too hard on a worm, it may break.
One part is in your hand.
The other part wriggles away.
How awful to tear an animal in two!
But the worm may not die.
You may have made two worms out of one.
The tail part may grow a new head.
The head part may grow a new tail.
Not many other kinds of animals can do this.

An earthworm feels soft and wet.

Its skin must be wet for the animal to live.

Like nearly all living things, a worm needs oxygen
to stay alive.

The worm takes in oxygen through its wet skin.

If a worm's skin dries out, the worm will die.

15

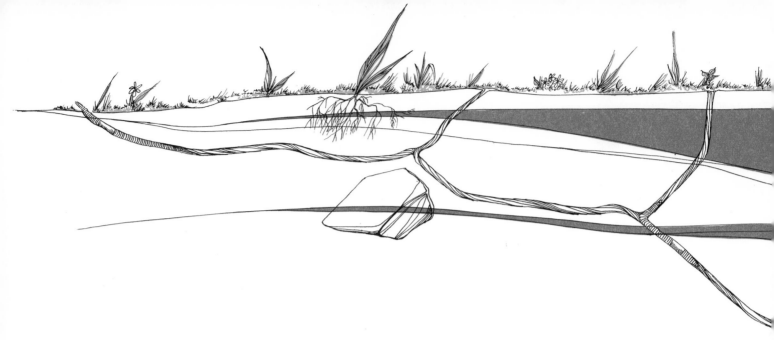

When the top of the soil is wet enough, earthworms
  will stay just a few inches below the surface.
As the soil dries in the summer, worms go deeper to
  find wet soil.
Some kinds of worms dig down six feet deep.
In wintertime earthworms also live deep underground, sometimes curled up in clumps.
There they sleep until the warmth of spring thaws
  the frozen ground above them.

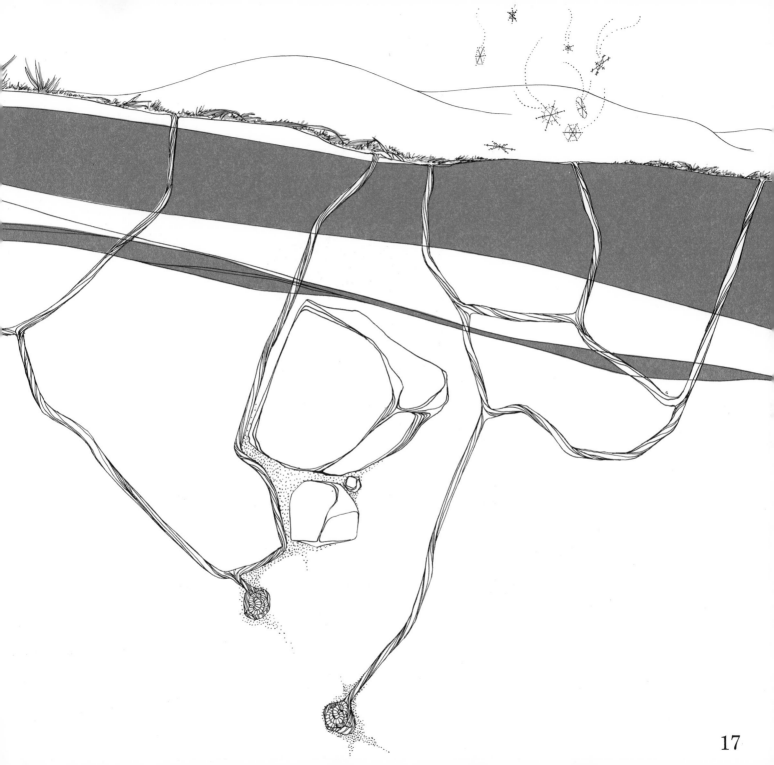

17

On a spring morning, after a heavy rain, you may
find earthworms crawling all over the ground.
People used to think that worms fell from the sky
when it rained.
But they do not fall with the rain.
They come out of the ground because their holes are
flooded.
Many of these worms die.
Scientists think they are killed by ultraviolet light
from the sun.
This is the same kind of light that gives some people
a sunburn.

Earthworms also come above ground to mate.

There are no male worms or female worms.

Each earthworm is both male and female.

Each produces sperm cells like a male and egg cells like a female.

Two worms lie side by side. A jellylike ring forms around them. Each worm gets sperm cells from the other. The sperm cells are placed in little hollows in the segments near the worms' heads. Each worm also develops egg cells inside its body. These are stored in other hollows. Then the worms separate. Each worm has half of the jellylike ring around its body.

Each worm backs up, making the slippery ring slide forward.

As it slides over the worm's segments, the ring picks up the egg and sperm cells.

Inside the ring, the sperm and egg cells join.

The eggs then begin to grow into young worms.

21

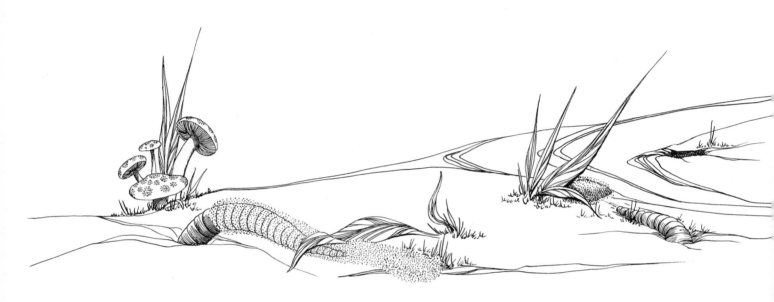

The ring slips over the worm's head like a sweater.
Then the ring folds together into a cocoon.
Some worms make cocoons the size of a pinhead.
  Some are as big as a grain of rice.
The cocoon is left lying in the soil.

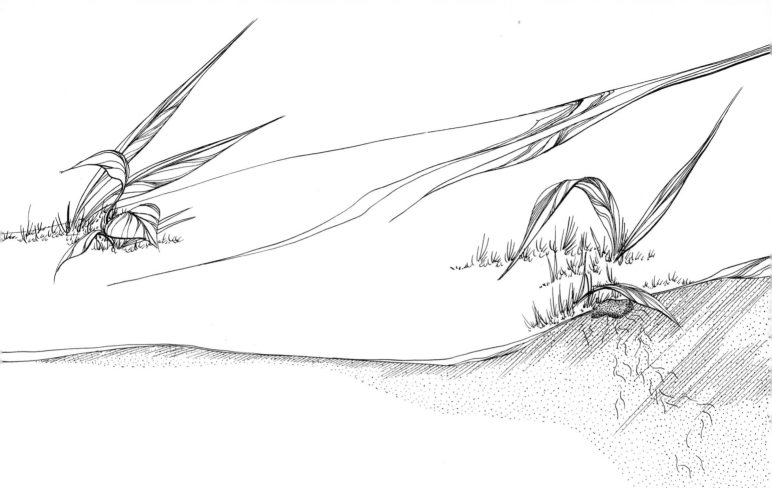

In two or three weeks, the tiny worms crawl out of
the cocoon and begin to dig down into the soil.
They look like bits of white thread.
In a few hours their skins turn dark.

Worms have many enemies.
In daytime they are hunted by robins and toads.
At night they are hunted by skunks and owls.

Moles, shrews, and fishermen hunt earthworms both
   day and night.
Even so, some earthworms may live for more than
   ten years.
But most worms probably live only a year or less.

More than 50,000 worms may live in a single big
   back yard.
All of these worms are moving through the soil.
They push their pointed heads through it and take
   bites of it.
Bits of leaves in the soil are the usual food for the
   worms.
But an earthworm will swallow anything that fits
   into its mouth, even tiny pebbles.

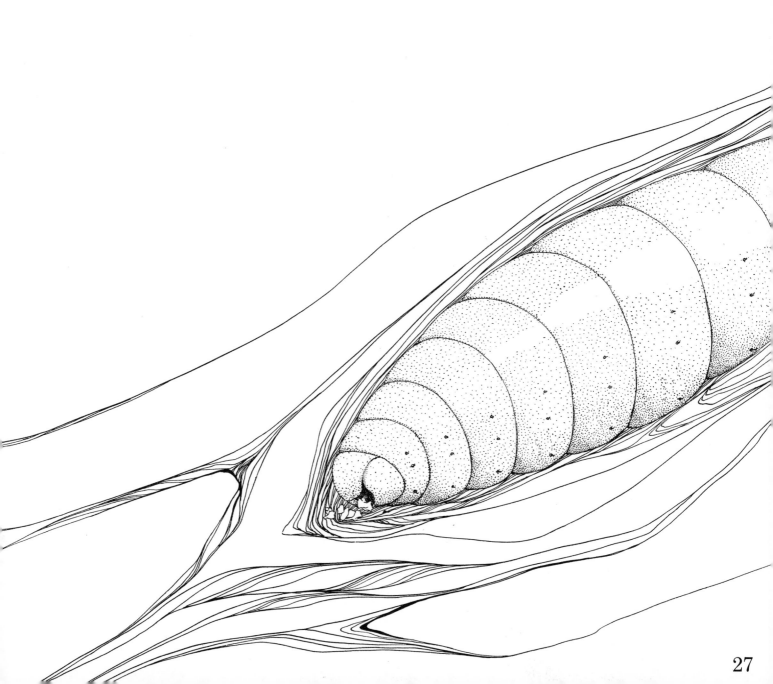

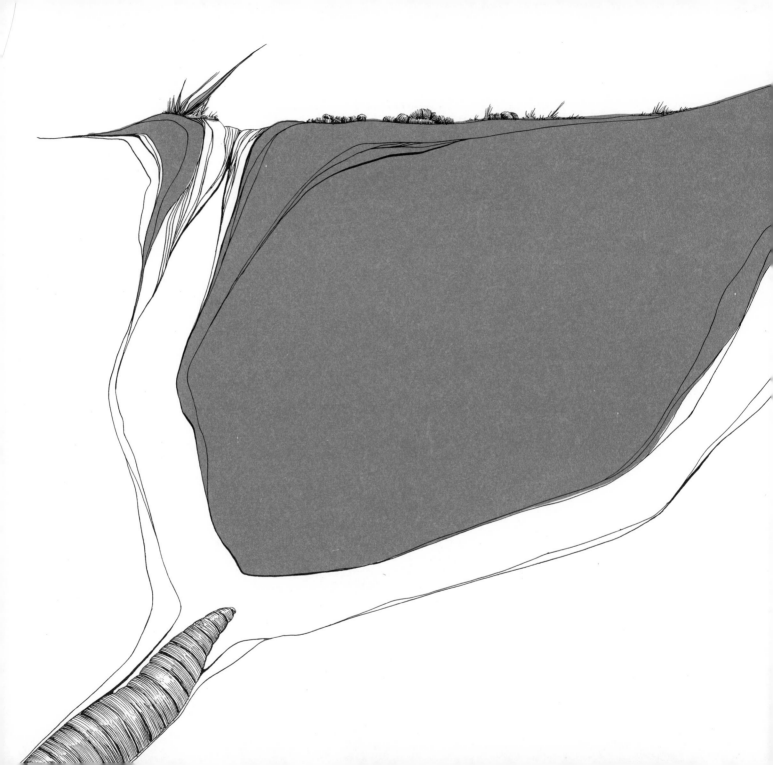

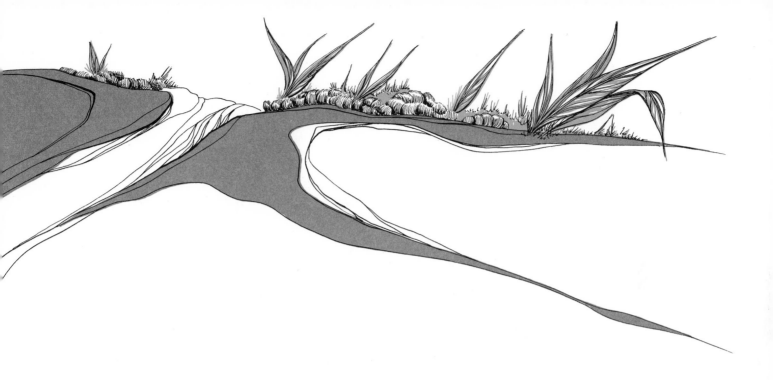

Some mornings you may see many little wriggly-
shaped piles of soil in a garden.

These dark mounds are called castings.

They are the body wastes of earthworms.

Castings are made up of soil, pebbles, and undigested
plant parts.

Some kinds of earthworms, including nightcrawlers,
leave their castings on the surface. Others leave
their castings underground.

A few worms by themselves do not have much effect on the soil.

But thousands of worms can.

In a year's time they mix and move many tons of soil.

Their castings are valuable for plants.

The castings contain minerals that plants need to grow well.

Earthworm burrows allow air and water to flow down easily into the soil.

This also helps plants to grow.

Earthworms are good for the soil and the plants that grow in it.

Perhaps the earth would not be such a beautiful planet if there were no quiet, crawling worms.

33

ABOUT THE AUTHOR

Laurence Pringle is the author of more than a dozen books whose subjects range from dinosaurs and their world to environmental problems today. His most recent book in the Let's-Read-and-Find-Out series was *Cockroaches: Here, There, and Everywhere.* Mr. Pringle has degrees in wildlife conservation from Cornell and the University of Massachusetts. One of his hobbies is nature photography, and many of his books are illustrated with his own photographs.

ABOUT THE ILLUSTRATOR

Peter Parnall's childhood was spent in several different parts of the United States—the Mojave Desert; Fort Davis, Texas; and New Haven, Connecticut. As a boy he kept many pets, and today his interest in the outdoors extends to the training of hawks and owls, and the breeding of a new species of game bird.

Mr. Parnall attended Cornell University and Pratt Institute. At various times he has worked as an air-hammer operator, a garage mechanic, a hand on a horse farm, and a tire salesman. He is now an advertising art director and a free-lance designer of packages and trademarks. Mr. Parnall has illustrated many children's books, a number of which have appeared on the *New York Times* list of best-illustrated books. He lives with his family in Milford, N. J.